Seed Production in Oil Palm:

A Manual

Techniques in Plantation Science Series

Series editors:

Brian P. Forster, Lead Scientist, Verdant Bioscience, Indonesia
Peter D.S. Caligari, Science Strategy Executive Director, Verdant Bioscience, Indonesia

About the series:

A series of manuals covering techniques in plantation science that form the essential underlying needs to carry out plantation science.

The series reflects the expertise in Verdant Bioscience that underlies the plantation science activities carried out at the Verdant Plantation Science Centre at Timbang Deli, Deli Serdang, North Sumatra, Indonesia.

Titles available:

1. **Crossing in Oil Palm: A Manual** – Umi Setiawati, Baihaqi Sitepu, Fazrin Nur, Brian P. Forster and Sylvester Dery
2. **Seed Production in Oil Palm: A Manual** – Eddy S. Kelanaputra, Stephen P.C. Nelson, Umi Setiawati, Baihaqi Sitepu, Fazrin Nur, Brian P. Forster and Abdul R. Purba
3. **Nursery Screening for *Ganoderma* Response in Oil Palm Seedlings: A Manual** – Miranti Rahmaningsih, Ike Virdiana, Syamsul Bahri, Yassier Anwar, Brian P. Forster and Frédéric Breton
4. **Mutation Breeding in Oil Palm: A Manual** – Fazrin Nur, Brian P. Forster, Samuel A. Osei, Samuel Amiteye, Jennifer Ciomas, Soeranto Hoeman and Ljupcho Jankuloski

Seed Production in Oil Palm:

A Manual

Eddy S. Kelanaputra
Verdant Bioscience, Indonesia

Stephen P.C. Nelson
Verdant Bioscience, Singapore

Umi Setiawati
Verdant Bioscience, Indonesia

Baihaqi Sitepu
Verdant Bioscience, Indonesia

Fazrin Nur
Verdant Bioscience, Indonesia

Brian P. Forster
Verdant Bioscience, Indonesia

Abdul R. Purba
Indonesian Oil Palm Research Institute, Indonesia

CABI is a trading name of CAB International

CABI	CABI
Nosworthy Way	745 Atlantic Avenue
Wallingford	8th Floor
Oxfordshire OX10 8DE	Boston, MA 02111
UK	USA
Tel: +44 (0)1491 832111	Tel: +1 (617)682-9015
Fax: +44 (0)1491 833508	E-mail: cabi-nao@cabi.org
E-mail: info@cabi.org	
Website: www.cabi.org	

A catalogue record for this book is available from the British Library, London, UK.

Library of Congress Cataloging-in-Publication Data

Names: Kelanaputra, Eddy S., author.
Title: Seed production in oil palm : a manual / Eddy S. Kelanaputra, Stephen
    P.C. Nelson, Umi Setiawati, Baihaqi Sitepu, Fazrin Nur, Brian P. Forster,
    Abdul R. Purba.
Description: Boston, MA : CABI, [2018] | Series: Techniques in plantation
    science series ; 2 | Includes bibliographical references and index.
Identifiers: LCCN 2018023338 (print) | LCCN 2018028695 (ebook) |
    ISBN 9781786395894 (ePDF) | ISBN 9781786395900 (ePub) |
    ISBN 9781786395887 (pbk : alk. paper)
Subjects: LCSH: Oil palm--Seeds--Production (Biology)
Classification: LCC SB299.P3 (ebook) | LCC SB299.P3 K45 2018 (print) |
    DDC 634.9/74--dc23
LC record available at https://lccn.loc.gov/2018023338

ISBN-13: 978 1 78639 588 7 (pbk)
        978 1 78639 589 4 (e-book)
        978 1 78639 590 0 (e-pub)

Commissioning editor: Rachael Russell / Rebecca Stubbs
Editorial assistant: Emma McCann
Production editor: James Bishop

Typeset by SPi, Pondicherry, India
Printed and bound in the UK by Severn, Gloucester.

Series Foreword – Techniques in Plantation Science

Verdant Bioscience, Singapore (VBS), is a new company established in October 2013 with a vision to develop high-yielding, high-quality planting material in oil palm and rubber through the application of sound practices based on scientific innovation in plant breeding. The approach is to fuse traditional breeding strategies with the latest methods in biotechnology. These techniques are integrated with expertise and the application of sustainable aspects of agronomy and crop protection, alongside information and imaging technology which not only find relevance in direct aspects of plantation practice but also in selection within the breeding programme. When high-yielding planting material is allied with efficient plantation practices, it leads to what may be termed 'intensive sustainable' production. At the same time, the quality of new products is refined to give more specialized uses alongside more commodity-based oil production, thus meeting the market demands of the modern world community, but with a minimal harmful footprint. An essential ingredient in all this is having sound and practical protocols and techniques to allow the realization of the strategies that are envisaged.

To achieve its aims, VBS acquired an Indonesian company called PT Timbang Deli Indonesia, with an estate of over 970 ha of land at Timbang Deli, Deli Serdang, North Sumatra, Indonesia, and the group works under the name of 'Verdant'. A central part of this estate, which will be used for important plant nurseries and field trials, is the development of the Verdant Plantation Science Centre (VPSC), to which the operational staff moved in October 2016. A seed production and marketing facility is now established at VPSC for commercial seed sales and the processing of seed from breeding programmes. The centre comprises specialized laboratories in cell biology, genomics, tissue culture, pollen, soil DNA, plant and soil nutrition, bunch and oil, agronomy and crop protection. Field facilities include extensive nurseries, seed gardens and trials (trial sites are also located at various locations across Indonesia). It is the aim of the company to use its existing and rapidly developing intellectual property (IP) to develop superior cultivars that not only have outstanding yield but also are resistant to both biotic and abiotic stresses, while at the same time meeting new market demands. Verdant not only develops and supplies superior planting materials but also supports its customers and growers with a package of services and advice in fertilizer recommendations and crop protection. This is all part of a central mission to promote green, eco-friendly agriculture.

Brian P. Forster and Peter D.S. Caligari
Lead Scientist and Science Strategy Executive Director
Verdant Bioscience

Contents

Acknowledgements

The authors are grateful to all the seed production and marketing, breeding and biotechnology teams of Verdant Bioscience for sharing their knowledge and providing helpful advice in preparing this manual.

Preface

As noted in the Foreword to this series, a central objective in Verdant Bioscience's mission is to sell superior varieties of oil palm, rubber and other plantation crops through plant breeding. Essential to this objective is the germination of seed from breeding programmes and for commercial production. Seed production is therefore central to Verdant Bioscience's activities and business success. Seed production methods have been developed to maximize germination, safeguard purity and provide high-quality planting material. The higher the quality of the planting material, both inherently (its genotype) and physiologically, the better the subsequent palm and its yield, thus helping sustainability by maximizing production per unit area of land. Protocols developed from these activities provide the basis for this manual. Our target audiences are students and researchers in agriculture, plant breeders, buyers, growers and end-users interested in the practicalities of oil palm seed production for breeding and sale.

Brian P. Forster and Peter D.S. Caligari
Series Editors
February 2018

Introduction

Abstract

Oil palm, *EIaeis guineensis* Jacq., is an important source of vegetable oil. Oil is extracted from the fruit mesocarp (crude palm oil) and from the seed (palm kernel oil). Botanically, the fruit is a drupe, with the kernel protected by a shell (nut). The commercial oil palm is Tenera (thin shelled), which is a hybrid from crossing Dura (thick shelled) seed palms with Pisifera (no shell) pollen palms. Since the shell is maternal tissue, seed for commercial planting has a thick shell. Like most seeds with very thick shells, oil palm seeds are difficult to germinate, the seed naturally germinates sporadically over time, and dormancy can last for up to 2 years. The challenges for seed production are to overcome dormancy (by weakening the operculum to allow germination), synchronize germination, produce a high germination percentage and high-quality germinated seed, free of abnormality or fungal infection. The processes used involve temperature treatments, imbibition, adjustment of seed moisture content and fungal control.

1.1   History of Oil Palm Production and Crop Facts

The African oil palm has the latin name *Elaeis guineensis* Jacq.: the genus name is derived from the Greek 'elaion', meaning oil, and the species name indicates its West African origin. The crop was discovered by travellers to Africa in the 15th century, but the first plantings in Indonesia, which led to its rise as the world's pre-eminent oil crop, did not occur until the late 19th century, with the first four Deli palms planted in 1848 (Pamin, 1998). Large-scale plantations were established in the early 20th century in both Africa and South-east Asia as interest in the crop developed. These initial plantations were composed of Dura palms, which are characterized as having thick-shelled fruits (Fig. 1.1). In the 1920s, the first crosses were made in deliberate

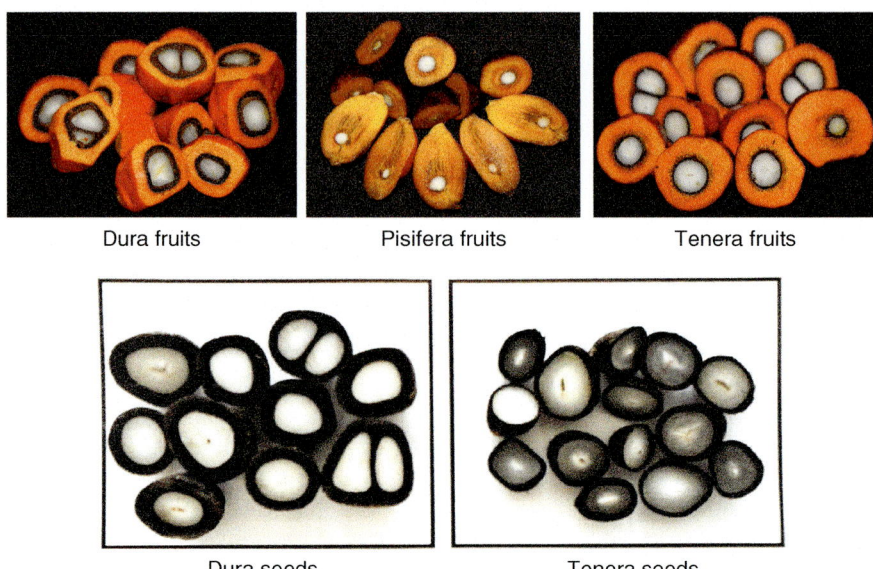

Dura fruits   Pisifera fruits   Tenera fruits

Dura seeds   Tenera seeds

Fig. 1.1. The appearance of Dura, Tenera and Pisifera fruits (sliced open) and Dura and Tenera seeds (nuts, thick and thin shelled, respectively).

attempts to improve the crop through plant breeding, and in the 1950–1960s, the more productive Tenera types (a result of crossing Dura with Pisifera; see Beirnaert and Vanderweyen, 1941) took over as the favoured commercial material in both Africa and South-east Asia. Tenera genotypes are thin shelled, have thick, oil-bearing fruit flesh (mesocarp) and yield 30% more oil than Duras. Thus, crossing became an essential and major component in commercial oil palm seed production, as well as in breeding.

Oil palm is grown in the humid tropics, usually between latitudes 10° north and south of the equator, and covers over 8.5 million hectares (Mha) worldwide. It is grown mainly from seed, although clonal plantings of tissue culture-produced ramets are also practised. The crop is highly profitable and grown both on large-scale plantations and by smallholders (Sayer *et al.*, 2012). Ripe, fresh fruit bunches are harvested continually, at intervals of 7–14 days, and sent to local mills for oil and kernel extraction. Oil palm fruits provide both crude palm oil (CPO) and palm kernel oil (PKO), extracted from the fruit flesh (mesocarp) and kernel (endosperm), respectively. CPO is made up of palmitic (43%), oleic (39%), stearic (5%) and other fatty acids (Siew, 2002), and is a major source of provitamin A and vitamin E (Barcelos *et al.*, 2015). PKO is a high-quality oil containing lauric (up to 50%), myristic (15%) and other essential fatty acids (Sambanthamurthi *et al.*, 2000). Since oil palm is harvested continually, CPO represents a relatively stable commodity compared to annual oil crops. The main CPO-producing countries are Indonesia (53% of global production) and Malaysia (38%); the largest consumers are India (28% of the market), Europe (22%) and China (22%).

1.2 Overview of Oil Palm Seed Production

Oil palm is the most popular oil crop in tropical regions. The first plantations of oil palm using commercially-produced seed were set up at the beginning of the 20th century. Malaysia developed the first oil palm seed market, launched in the 1970s, which included very strict planting programmes. Indonesia has the largest oil palm plantation area in the world and is the biggest CPO producer and the largest seed producer in the world.

Table 1.1. Oil palm D×P seed production capacity worldwide 2009. (From Khusairi *et al.*, 2010.)

Country	Seed production (million)
Asia	
Indonesia	250
Malaysia	81.5
Papua New Guinea	30
Thailand	13
India	2
Total	376.5
Central-South America	
Costa Rica	30
Honduras	2
Colombia	2
La Cabana	1.5
Ecuador	2
Brazil	1
Total	38.5
Africa	
Benin	6
Nigeria	2
Cameroon	2
Ghana	2
Democratic Republic of Congo	3
Ivory Coast	10
Total	25
Grand total	440

1.3   Biology and Genetics of Oil Palm

Oil palm (*E. guineensis*) is a long-lived perennial. It has a single apical meristem and does not sucker/conventionally reproduce asexually. After germination, there is a 3-year juvenile stage before inflorescences appear. These are produced in each leaf axil and are either male or female (monoecious), the first inflorescences produced after the juvenile growth phase (2 years) are normally male. Oil palm is an outbreeding species and pollination is affected predominantly by the weevil, *Elaeidobius kamerunicus*, which depends on oil palm inflorescences to complete its life cycle. Oil palm inflorescences are large and typically give rise to bunches containing 100–4000 fruits, depending on palm age. Fruits mature at about 150 days after pollination, turning from black to red, and the outer fruit begins to fall out/abscise from the bunch when it is ripe. The fruit is a drupe (stone fruit) composed of a fleshy mesocarp and a central kernel protected by a shell (endocarp). The kernel contains the products of fertilization: embryo and endosperm. In the wild, the mesocarp of fallen fruits rots or is eaten, leaving behind the kernel: germination takes place in favourable conditions by the emergence of a seedling shoot and root through the germ pore in the shell. Generally, one seedling is produced per seed, but up to three may occur.

Oil palm is diploid, with 16 pairs of chromosomes, but is thought to have evolved from an ancient tetraploid species as there is extensive genome duplication (Singh *et al.*, 2013). It is highly heterozygous owing to its outbreeding reproductive system. Although oil palm can be artificially self-pollinated, inbreeding depression has been widely reported. Genetic maps of the oil palm genome have been developed (Mayes *et al.*, 2000; Billotte *et al.*, 2005) and the genome has been sequenced (Singh *et al.*, 2013); thus, genetic markers may be deployed in screening for genes of interest in progeny from deliberate crossings.

1.4   Crossing and Seed Type

Commercial seed production is based on Dura (D) × Pisifera (P) crosses, which produce the desired thin-shelled Tenera (T). The predominant parental lines have been Deli Duras and AVROS Pisiferas. In addition to commercial seed production, seed needs to be germinated from breeding programmes in developing new varieties. Crop improvement through breeding has been limited by the genetic variation contained in elite Dura and Pisifera parental gene pools. In order to make progress in breeding, it became necessary to provide breeders with more genetic variation, and thus germplasm collections from wild, landrace and cultivated materials in West Africa (the centre of diversity) have been carried out. The West African Institute for Oil Palm Research (WAIFOR), Nigeria, was one of the first to do this. In recent years, major oil palm breeding companies have joined collecting expeditions in West Africa (Okyere-Boateng *et al.*, 2008; Sapey *et al.*, 2012). Crossing programmes include D×P, D×D, D×T, T×T, T×D and T×P crosses (Forster *et al.*, 2018; Setiawati *et al.*, 2018), and germination for both thick- (D female) and thin- (T female) shelled seed is required.

1.5 Commercial Crossing

Tenera oil palm has become the predominant commercial fruit type. Tenera seed is produced by commercial seed companies by crossing Dura (thick shelled) with Pisifera (no shell) genotypes. Pisifera palms are often female sterile and do not set fruit. Pisiferas are therefore used as the male (pollen) parents in commercial production (Setiawati *et al.*, 2018). There is no reason why fertile Pisiferas should not also be used as pollen parents, but there was a concern that the resulting Tenera progeny had a thicker shell (Menendez and Blaak, 1964).

Pisifera pollen is collected and stored ready for use in crossing. The shell thickness trait is controlled by a single gene, *Sh*, with Dura being homozygous thick shell (*Sh/Sh*), Pisifera being homozygous no-shell (*sh/sh*) and Tenera being heterozygous (*Sh/sh*). Seed production is a specialized and lucrative business; in 2014, Indonesian oil palm seed producers sold 102,826,918 seeds at an average cost of about US$0.8/seed. The oil palm industry is therefore dependent on quality-controlled crossing procedures and seed production.

1.6 Seed Production

The term 'seed production' in oil palm for breeding and commercial production refers to germinated seed, not biological seed. Seed (botanically a nut) is sold (commercial) and planted in nurseries (commercial and breeding materials) as germinated seed (Fig. 1.2).

Fig. 1.2. Germinated seed ready for sale and planting.

The process for both commercial and breeding seed production starts with deliberate and controlled crossing of parental lines, followed by bunch harvesting, removal of the fruit mesocarp and the delivery of fresh, ungerminated seed to a seed production facility. These and subsequent seed germination protocols are described in the following chapters.

References

Barcelos, E., de Almeida Rios, S., Cunha, R.N., Lopez, R., Motoike, S.Y., Babiychuk, E., *et al.* (2015) Oil palm natural diversity and the potential for yield improvement. *Frontiers in Plant Science* 6, 190.

Beirnaert, A. and Vanderweyen, R. (1941) Contribution a l'étude génétique et biométrique des variétiés *d'Elaeis guineensis* Jacquin. *Institut National pour l'Etude Agronomique du Congo belge (INÉAC)* 27, 1–101.

Billotte, N., Marseillac, N., Risterucci, A.M., Adon, B., Brottier, P., Baurens, F.C., *et al.* (2005) Microsatellite-based high-density linkage map in oil palm (*Elaeis guineensis* Jacq.). *Theoretical and Applied Genetics* 110, 754–765.

Forster, B.P., Sitepu, B., Setiawati, U., Kelanaputra, E.S., Nur, F., Rusfiandi, H., *et al.* (2018) Oil palm (*Elaeis guineensis*). In: Campos, H. and Caligari, P.D.S. (eds) *Genetic Improvement of Tropical Crops.* Springer, Cham, Switzerland, Chapter 8, pp. 241–290.

Khusairi, A., Zamzui, I., Ong-Abdullah, M., Samsul Kamal, R., Ooi, S.E. and Rajanaidu, N. (2010) Production, Performance and Advance in Oil Palm Tissue Culture. *A paper presented at the International Seminar on Advances in Oil Palm Tissue Culture*, held on 29 May 2010 in Yogyakarta, Indonesia. Malaysian Palm Oil Board, Selangor, Malaysia, pp. 1–23.

Mayes, S., Jack, P.L. and Corley, R.H.V. (2000) The use of molecular markers to investigate the genetic structure of an oil palm breeding programme. *Heredity* 85, 288–293.

Menendez, T. and Blaak, G. (1964) Plant breeding division. In: *12th Annual Report.* West African Institute Oil Palm Research, Benin, Nigeria, pp. 49–75.

Okyere-Boateng, G., Dwarko, D.A., Kaledzi, P.D. and Nuertey, B.N. (2008) Collection, conservation and evaluation of the disappearing oil palm (*Elaeis guineensis* J) landraces in Ghana. *International Journal of Pure and Applied Sciences* 1, 18–31.

Pamin, K. (1998) A hundred and fifty years of oil palm development in Indonesia: from the Bogor Botanical Garden to the industry. In: Pamin, K. (ed.) *Proceedings 1998 International Oil Palm Conference 'Commodity of the past, today and the future'.* Indonesian Oil Palm Research Institute, Medan, Indonesia, pp. 3–23.

Sambanthamurthi, R., Kalyana, S. and Tan, Y. (2000) Chemistry and biochemistry of palm oil. *Progress in Lipid Research* 39, 507–558.

Sapey, E., Adusei-Fosu, K., Agyei-Dwarko, D., Okyere-Boateng, G. and Bois D'Enghlen, D. (2012) Collection of oil palm (*Elaeis guineensis* Jacq) germplasm in the northern region of Ghana. *Asian Journal of Agricultural Sciences* 4, 325–328.

Sayer, J., Ghazoul, J., Nelson, P. and Boedhihartono, A.K. (2012) Oil palm expansion transforms tropical landscapes and livelihoods. *Global Food Security* 1, 114–119.

Setiawati, U., Sitepu, B., Nur, F., Forster, B.P. and Dery, S. (2018) *Crossing in Oil Palm: A Manual. Techniques in Plantation Science*. Forster, B.P. and Caligari, P.D.S. (eds). CAB International, Wallingford, UK (in press).

Siew, W.L. (2002) Palm oil. In: Gunstone, F.D. (ed.) *Vegetable Oil in Food Technology: Composition, Properties and Uses*. Wiley-Blackwell, Hoboken, New Jersey, pp. 25–58.

Singh, R., Ong-Abdullah, M., Low, E.T.L., Arif, M., Manaf, A., Rosli, R., *et al.* (2013) Oil palm genome sequence reveals divergence of interfertile species in old and new worlds. *Nature* 500, 335–339.

Health and Safety Considerations   2

Abstract

All field and laboratory operations should have standard health and safety protocols. These may vary according to local requirements and standards. Some equipment will also come with instructions on proper use, which may involve training, including health and safety issues. Failure to abide by these can result in accidents and personal injury (serious and minor); neglect of health and safety issues may incur penalties such as fines or cessation in field and laboratory activities. Guidelines in health and safety issues relating to crossing in oil palm are given below.

2.1 Health and Safety in the Field

Oil palm seed production is carried out in the field, and a major health and safety issue is the height of palm trees. Inflorescences are located in the canopy of the palm trees, and these can reach a height of 15 m after 20 years. It is recommended that young, short palms are used whenever possible, but if older, taller trees are used for either pollen collection and/or crossing, then several safety issues must be deployed.

Equipment needed for tall palms:

- Ladders or scaffolding – used for climbing tall trees.
- Mechanized elevators (cherry pickers) – used for tall palms or any palm where there is concern about the integrity of the palm stem, e.g. risk of *Ganoderma/Fusarium* disease.
- Harnesses, helmets, boots, etc. – used when working on tall palms.
- Clothing in general.

Other general considerations:

- Sharp knives.
- Dangerous insects or animals, such as mosquitoes and snakes.

- Standard operating procedures (SOPs) and training.
- Working alone.
- Emergency procedures, first aid box.
- Wear a mask: hazardous chemicals (formalin and insecticides) are used (sprayed) in the isolation of female inflorescence to safeguard against uncontrolled pollinations. Face protection is therefore required to prevent inhalation and eye and skin contact.
- Wear gloves: hazardous chemicals (formalin and insecticides) are used (sprayed) in the isolation of female inflorescence to safeguard against uncontrolled pollinations. Hand protection is therefore required to prevent skin contact. Any contact with hazardous chemicals should be followed immediately by washing and first aid.

2.2   Health and Safety in the Seed Processing Facility

Oil palm seed production requires various processing procedures, especially in treatment, storage and seed preparation. Good processing practices are therefore required.

- Use safety shoes when you are in the seed preparation working area; remove and change to clean sandals before entering the processing area; and remove when leaving the processing area. This provides protection to yourself and the samples you are working with and protects people outside the area from contamination by processing materials.
- Wear eye protection during the seed treatment process when working with chemicals, then remove before leaving the area.
- Wear gloves, goggles, mask, boots and apron to protect yourself from hazardous chemicals such as fungicides and to protect samples from fungal contamination. Hand-wash sinks are required for protecting workers from exposure to chemicals and seeds that have been treated with fungicides. Gloves should be worn when handling chemicals. Be aware of other workers in the treatment area and avoid touching commonly used utilities such as light switches, door handles, taps, telephones, etc., while wearing gloves. Remove gloves before leaving the treatment area.
- Training should be provided for treatment processes, to avoid any incident during the treatment process.
- Be aware of emergency procedures: fire-fighting, emergency exits, emergency telephone numbers, and the location of fire extinguishers, eyewash and first aid/first aiders.
- Be aware of hazards relating to chemicals used in the laboratory and read their Material Safety Data Sheet (MSDA information is available on the Internet), which provides information on health and safety, first aid, fire and explosion risks, disposal, how to clean up spillage, handling and storage.

- Be aware of SOPs that have been developed for your working area, or which should be developed (for example, for waste disposal).

More information on safe procedures in the laboratory are given by Barker (2005).

Reference

Barker, K. (2005) *At the Bench: A Laboratory Navigator*. Cold Spring Harbor Press, New York.

Isolation of the Female Inflorescence

3

Abstract

Care is required when isolating female inflorescences to prevent damage. The developing inflorescence should be treated with insecticides to prevent insects, particularly the oil palm-pollinating weevil, from leading to pollen cross-contamination. Specialized isolation bags are designed with a window to inspect and monitor development and through which subsequent operations (spraying and pollination) are performed. The isolation process involves insect-proofing as insects, particularly weevils, carry unwanted pollen and cause seed contamination. Instructions are given on the procedures used in isolating the female inflorescence that maximize success in artificial crossing.

The procedures involve the use of hazardous chemicals (insecticides) and may involve climbing tall palms. Guidance in health and safety procedures are given in Chapter 2 of this manual.

3.1 Tools

- Knife – used for cleaning the female flower to be isolated.
- Cotton wadding – used to put insecticides (Furadan) on the stalk to prevent insects from getting in.
- Terylene bags – used for isolating the female flower.
- Rubber band – used to tie up the terylene bag at the stalk.
- Hand sprayer – used for spraying a formalin solution.

3.2 Reagents

- 2% formalin solution – used to sterilize flowers from other pollen.
- Insecticide such as Furadan 3G and aerosol spray – used to kill insects that can bring other pollen into the terylene bag and contaminate the pollination.

© Eddy S. Kelanaputra, Stephen P.C. Nelson, Umi Setiawati, Baihaqi Sitepu, Fazrin Nur, Brian P. Forster and Abdul Razak Purba 2018. *Seed Production in Oil Palm: A Manual*   15

3.3   Methods

Step 1

Palms are inspected for the emergence of female flower bunches. These are noted and their development is monitored prior to isolation.

Selected female
inflorescence for isolation

Step 2

The outer and inner spathes are removed from the selected female inflorescence using a knife, to expose the developing florets. The inflorescence is then sprayed with 2% formalin to kill any pollen and insects present. A wad of cotton wool containing insecticide is then tied to the inflorescence stalk using rubber bands, this kills any insects that may carry unwanted pollen.

Cut to open the spathes

Spray with 2% formalin

Isolate female
inflorescence with
pollination bag

A wad of cotton wool placed around the stalk	Tied with a rubber band

Step 3

After the removal of the spathes, the female inflorescence is covered with two terylene bags. The inner bag is a new bag, the outer bag is normally a recycled old bag. The bags are equipped with a window for monitoring development and through which pollination is effected. At least 9 days are needed for the female florets to become receptive, fully opened flowers. If the female inflorescence becomes receptive in less than 9 days, it is rejected, as uncontrolled pollination may have occurred.

Outer layer with a used bag	Tied with a rubber band	Record the palm identity and isolation date

Pollination

4

Abstract

Pollination is controlled through a window in the specialized isolation bag; this reduces the risk of contamination (foreign pollen, disease and insects). Prior to pollination, the area around the isolation bag is sprayed to kill insects, as these can carry unwanted pollen. Pollen is normally mixed and diluted with talcum powder to provide an inert spreading media. This mixture is blown on to receptive female inflorescences using a surface-sterilized 'popper' that has a tube containing silica gel to adsorb the moisture from the pollinator's breath. The isolation bag is also shaken to spread the pollen over the entire inflorescence. Labels are then attached showing pollination dates and parental genotypes. The isolation bags may be removed once fruit set is established and harvested when fruits are mature.

4.1   Pollination

Pollination is normally carried out in the morning when nectar is visible on female florets. Mature pollen grains of oil palm are binucleate, containing one vegetative and one generative nucleus. When the pollen grain lands on a receptive stigma of a female inflorescence, it begins to germinate. The pollen tube grows down the style, carrying with it the two nuclei; the generative nucleus divides to produce two sperm cells, which are deposited into the ovary, where they effect fertilization. For more details on oil palm pollen development, see Nasution *et al.* (2009).

4.2   Quality Control: Blank Pollination

Blank pollination is performed randomly using talcum powder to monitor the quality of each pollinator's work and to assess whether there is any contamination that will result in seed being set. Pollinators are not aware which

© Eddy S. Kelanaputra, Stephen P.C. Nelson, Umi Setiawati, Baihaqi Sitepu, Fazrin Nur, Brian P. Forster and Abdul Razak Purba 2018. *Seed Production in Oil Palm: A Manual*

samples are pure talc (blank) or which are normal talc mixed with pollen. Dura contamination is a concern, as it reduces seed quality. The Indonesian government standard for this is a maximum of 2% Dura contamination, and seed producers generally show their standard as far less than this (normally less than 0.1%). Contamination results in the development of fruits that are checked and counted 3 months after blank pollination. These are then scored as a percentage of the total number of flowers. The work of the pollinator must be tested every month by giving him or her blank pollen to pollinate the inflorescence flower. Premiums are often given to pollinators who pass the test with less than 0.1% seed set, and penalties/increased supervision/training given to those who fail to achieve less than 0.1% seed set in 3 months.

4.3   Tools

- Pollen mixing box – used for mixing pollen in sterile conditions.
- Pollination tube (popper, pollen blower) – used to blow pollen over the inflorescence.
- Labels and wire – used to identify the pollination.
- Net bag – used for wrapping the pollinated bunch.
- Scissors are used for cutting the rubber band and for piercing a hole in the window of the inflorescence bag.

All tools used for pollination are sterilized using 96% ethanol and heating.

| Spray with 96% alcohol | Heat the box to min temperature of 100°C for 10 min |

4.4   Reagents

- 96% ethanol – used to surface sterilize equipment.
- Insecticide, such as an aerosol spray – used to kill insects that may enter the inflorescence.
- Talcum powder – used to mix with pollen to act as a carrier in providing even distribution of pollen when blown over the inflorescence.

4.5 Methods

Step 1

Prior to pollination, the area around the isolated female inflorescence is sprayed with insecticide to kill insects that may carry unwanted pollen.

| Spray around the isolated bunch with insecticide | Cut the top side of the bag to open the inner bag | Spray inside the outer bag with insecticide |

Step 2

Pollen for pollination is mixed with talcum powder, usually in a ratio of 0.03–0.20 g of pollen (depending on pollen viability, pollen usually has >80% viability) with 1 g of talcum powder. Pollen mixing is carried out in a sterilized pollen mixing box to prevent contamination.

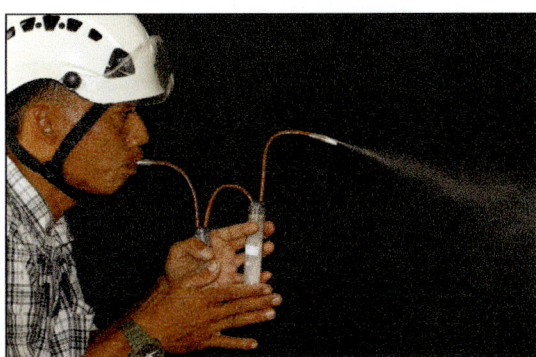

| Pollen blowing using popper |

Step 3

Open the outer isolation bag by cutting the top side of the bag. Pollination is effected by blowing the pollen plus talc mixture into the inner isolation bag, by piercing a hole in the bag window using scissors (when preparing the pollen blower equipment, make sure the hole is sealed with insulation tape). Pollination is normally carried out twice, on two consecutive mornings, to achieve maximum percentage fruit set. The isolation bag is shaken after each delivery of pollen by the pollinator to help distribute the pollen over the whole inflorescence. The bag is resealed with the isolation tape and labelled with cross details (female and male parents and dates). Then cover the inner bag with the outer bag and re-close the outer bag by tying it along the cut side. This method also allows supervisors to see which inflorescences have been pollinated.

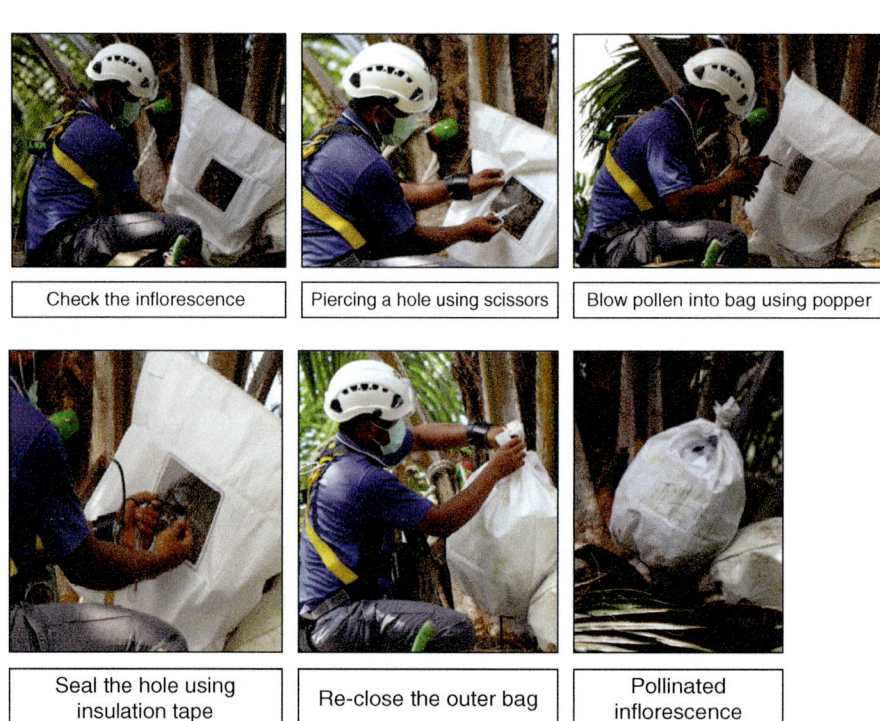

| Check the inflorescence | Piercing a hole using scissors | Blow pollen into bag using popper |

| Seal the hole using insulation tape | Re-close the outer bag | Pollinated inflorescence |

Step 4

At 21–25 days post-pollination, the isolation bags (inner and outer) are removed and the bunch is labelled by inserting a wire with a label into the developing bunch.

Label the bunch

Step 5

For the purpose of the blank pollination test, the bunch is inspected for fruit set on 60–67 days post-pollination.

Blank pollination (no seed) bunch   Spikelet (no seed)

Reference

Nasution, O., Rusfiandi, H., Sitorus, A.C., Forster, B.P., Nelson, S.P.C. and Caligari, P.D.S. (2009) Cytological studies of pollen development in oil palm (*Elaeis guineensis* Jacq.). In: *Proceedings of Agriculture, Biotechnology and Sustainability Conference*, PIPOC 9–12 November 2009, Kuala Lumpur, Malaysia, Vol 2. Malaysian Palm Oil Board, Kuala Lumpur, pp. 954–961.

Harvesting

5

Abstract

The ripe bunch with red-coloured fruits can be harvested about 150 days after pollination, or when one loose fruit appears on the ground. Harvested bunches are sent to the processing area in gunny sacks, along with any loose fruit that detach during harvesting. Minimum ripeness standards are used to reduce the risk of loose fruit/seed being lost, and the use of net bags after bag removal at 21–25 days post-pollination will prevent any loose fruit loss.

5.1 Tools

- Chisel – used for cutting the fruit bunch from the tree.
- Net bag – used to wrap the bunch after pollination to avoid any missing loose fruits when it is harvested.
- Gunny sack – used to put the bunch inside for transporting to the seed processing building.
- Labels – used to record the crossing identity and date of pollination.
- Bunch transporter – used to transport the bunch from the field to the seed processing area.

5.2 Methods

Step 1

Cut the fruit bunch stalk using a chisel, care being taken to avoid fruit damage. The bunch will fall to the ground once detached and may shed more loose fruits. Check that the label is still attached to the bunch.

Step 2

Check the net bag is still fit for purpose and that no fruits have escaped after felling the bunch, then put the bunch with the net bag into a gunny sack.

Step 3

Put the gunny sack into a transporter, a three-wheeled motorbike or truck for example, and send to the processing building.

Seed Preparation

6

Abstract

In order to collect fresh seeds from harvested bunches, the fresh bunch is chopped using an axe (a trained worker must be assigned to this job) to separate the spikelets from the stalk. Fruits attached to spikelets are fermented for 3–4 days in an area separate from the detached fruits. Detached fruits are peeled by hand by a skilled person using a sharp knife. Detached fruits are placed in a de-pulping machine to remove the mesocarp from the seed (nut). Seeds that still have fibre are cleaned by scraping with a sharp knife; this reduces fungal contamination. The clean fresh seeds are then treated with a disinfectant and fungicide solution for about 3 min and placed into perforated trays to dry in the open air under a fan for about 24 h. The dried seeds are sorted to remove off-types (white, small and broken) and transferred to slow drying conditions at 19–23°C for about 24 h.

6.1 Tools

- Sharp knives – used to detach fruits from their spikelet.
- Axe – used to chop the spikelet from the bunch.
- De-pulping machine (often referred to as a depericarper) – used to remove the mesocarp (fruit pulp) from the seeds.
- Perforated plastic basket – used as a seed container and carrier; a bunch reference label is placed inside the basket.
- Weighing scales – used to measure the dosage of fungicide.
- Measurement glass – used to measure the dosage of Chlorox and Teepol.

Safety equipment for seed preparation

Goggles masker, leather hand gloves, helmet, boots, ear plugs and fire extinguisher. Note, the floor around the de-pulper is usually slippery, due to oil from the fruits, and therefore protective clothing is required (including a helmet) and care should be taken.

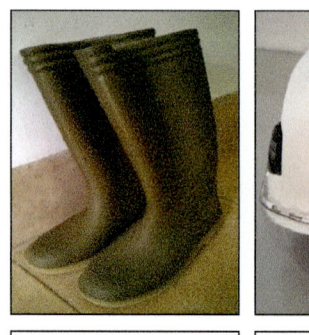

| Boots | Helmet | Fire extinguisher |

6.2   Materials

- Growth regulator Ethephon.
- Fungicides and surfactant.

6.3   Methods

Step 1

Check the fruit bunch and label based on the harvesting list. Weigh the bunch and place on to a chopping table. Chopping is carried out to separate the spikelets from the main bunch stalk; the spikelets are then placed into a basket to be 'fermented'. Alternatively, inject the stalk of the bunch with Ethephon (an ethylene producer) and store for 1 day in gunny bags in a dry, ventilated room at ambient temperature. After storage, the bunch can be chopped with greater ease and the fruits will detach from the spikelets.

| Bunch chopping | Spikelet storage before fruit separation |

Step 2

Peel the fruits from the spikelets and place into a de-pulping machine (often known as a de-pericarper, although the shell (endocarp) is not removed) to remove the mesocarp and leave the seeds (nuts). Normally, the speed of the de-pulper is 250 rpm and the process should be stopped when the water output (waste) from the de-pulper runs clean (which takes about 15 min). If there is any remaining fibre on the seeds, this can be removed by scraping with a sharp knife.

| De-pulper | Seed scraping |

Step 3

Clean the seeds with running water and treat with 15 ml Teepol in 10 l of water; then clean them again with running water and soak in a solution of 350 ml of Chlorox (5%) in 7 l of water. The next treatment is to soak the seeds in fungicide in 10 l of water for 3 min. Then dry the seeds for 6 h under a fan at ambient temperature.

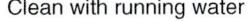
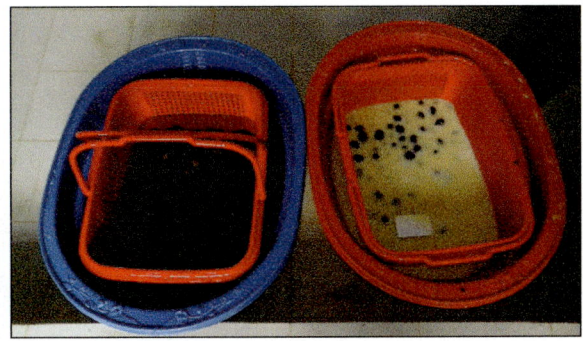

| Clean with running water | Seed treatment: dip into fungicide solution |

Step 4

Remove any off-type seeds (such as tiny, broken or white seeds), then count the good seeds before transferring them to a slow-drying room in perforated trays (stacked) at 19–23°C. After approximately 1 day, check the seed moisture in a sample of seed. If the moisture content meets requirements (17–20%), the seeds can be moved into plastic trays or plastic bags and stored in a cold room at 19–23°C, i.e. the same temperature as slow drying. Seeds may be stored in the cold room usually for up to 1 year without a major drop in seed viability.

| Rapid air drying | Slow drying |

Seed Viability Testing

7

Abstract

Seed viability testing in seed processing is carried out to forecast germination levels. The test is destructive and involves removing the embryo from the kernel for visual inspection.

7.1 Tools

- Petri dish – used for embryo observation.
- Tissue – used to place embryos into the Petri dish.
- Hammer – used to crack open seed (nut).
- Plastic bag – used to place seed for cracking so that broken shell is contained.
- Scalpel – used to slice through the kernel (endosperm) to expose and lift out the embryo.

7.2 Methods

Step 1

The first step is to take a seed sample from a batch being processed that is representative of the batch.

Step 2

Sampled seeds are placed into a plastic bag and cracked carefully using a hammer to avoid embryo damage.

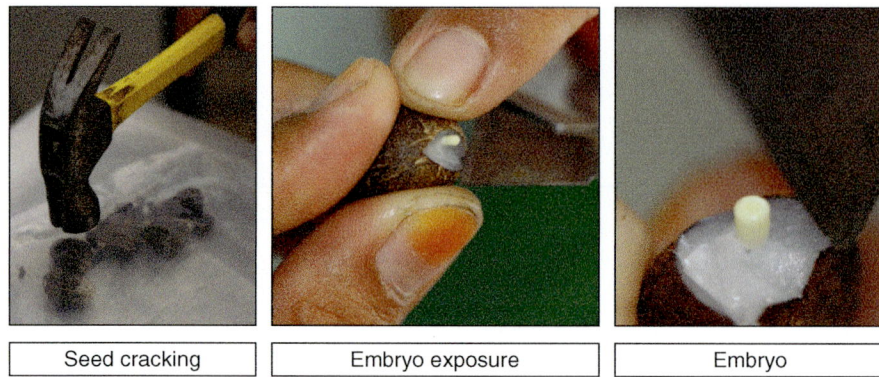

| Seed cracking | Embryo exposure | Embryo |

Step 3

Embryos are removed from the kernel with a scalpel and placed on to wet tissue in a Petri dish. The embryos are arranged based on bunch reference and checked visually for viability: viable embryos are torpedo-shaped and pale yellow; dead embryos are shrivelled and dark yellow/brown. The percentage of viable embryos is calculated by dividing the number of healthy embryos by the total number of embryos × 100%.

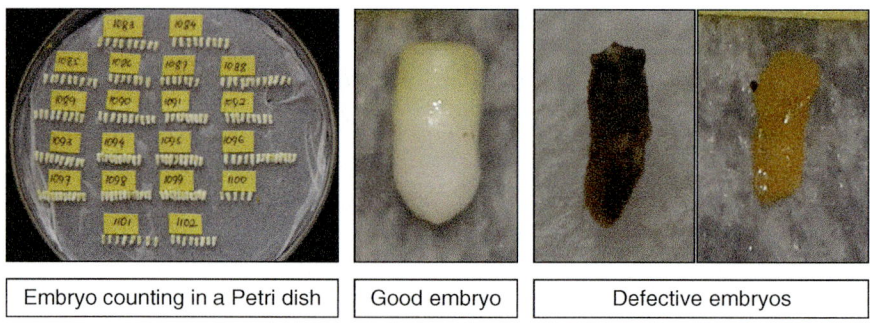

| Embryo counting in a Petri dish | Good embryo | Defective embryos |

Seed Moisture Testing 8

Abstract

Seed moisture content is critically important for efficient seed processing. In order to achieve the highest seed germination, the moisture must be controlled: if it is too low, the seed will die; and if too high, it will stimulate fungal growth and result in seed loss during processing.

8.1 Tools

- Oven – used to dry seed.
- Balance – used to weigh wet and dry seeds.
- Glass beaker – used to contain seeds for oven drying.
- Hammer – used to crack open seeds to separate the kernel from the shell.
- Plastic bag – used to contain cracked seed parts.
- Cutter – used to separate the embryo from the kernel.

8.2 Methods

Step 1

The first step is to take seed samples from a representative sample of seed bunches being processed.

| Sample taken from drying rack | Collected samples |

Step 2

Weigh the seed sample using a balance and record, then place the sampled seeds into a plastic bag and crack the shells carefully using a hammer.

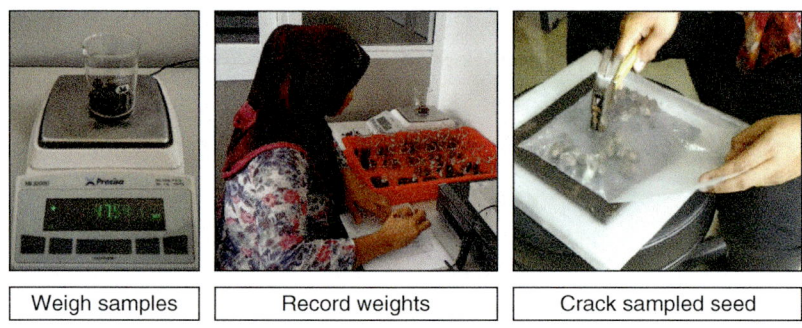

| Weigh samples | Record weights | Crack sampled seed |

Step 3

Separate the shell and kernel and weigh each separately; record as wet weights (WW).

| Weigh shell samples | Weigh kernel samples |

Step 4

Place the cracked shell and kernel samples into a drying oven at 80–110°C for 24 h. Reweigh the dried shell and kernel samples separately and record as dry weight 1 (DW1). After cooling for 1 h, the sample is re-dried in the oven and the samples reweighed; this is recorded as dry weight 2 (DW2). DW2 is taken to ensure there is no significant difference in seed moisture before and after cooling.

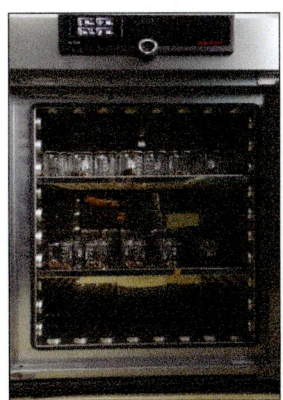  

| Dry sample in oven | Dry sample weighing |

Step 5

The percentage moisture content of the seed is calculated by subtracting the dry weight from the wet weight and dividing by the dry weight, then multiply by 100%: (WW – DW1)/DW1 × 100%. After cooling down, seed moisture 2 = (WW – DW2)/DW2 × 100%.

Seed Processing for Commercial Production

9

Abstract

Oil palm seeds have sessile embryos and, for germination to start, the operculum that blocks the germ pore (and essentially makes the seed dormant) has to be weakened. Seed processing procedures must be implemented and monitored to ensure maximum and uniform germinations are achieved; the usual commercial target is over 80%.

9.1   General Principles in Commercial Seed Production

Minimum seed stock

For commercial seed production, it is recommended to maintain a seed stock in a cold room (for 2–3 months), to allow the facility to be responsive to new seed orders.

Maintain seed quality standards

Avoid and discard seeds damaged during the de-pulping process and remove off-type seeds.

No fungal contamination

The seed storage and processing areas must be kept clean at all times and seed moisture levels monitored at all times.

Excellent germination and low rejection rates

This requires strict implementation of seed production procedures. Optimum temperatures in a hot room and the germination rooms must be maintained with manageable seed tray stacking. Seed disturbance and seed sorting should be minimized during germination to avoid abnormal germination.

Heating is the common treatment used to weaken the operculum so that seed germination will start. Temperature control and the duration of the heating period is critical and needs careful attention. Moisture and oxygen uptake are also essential factors in the control of germination. Soaking, drying and periodic seed spraying with distilled water are required to achieve the target seed moisture contents. Good air circulation with optimum temperatures during processing will also have a significant impact on seed germination.

In nature, wild oil palm seeds respond to heat produced from the exo-thermic decomposition of the oily mesocarp by fungi and other microorganisms (Alang *et al.*, 1988). This natural process is very slow and relatively few seeds germinate. High fungal infection will cause seeds to fail; thus, periodic seed inspections for fungal infection are needed during processing.

Germinated seeds with standard radicle (primary root of the embryo) and plumule (primary shoot of the embryo) lengths are selected and collected. The maximum radicle and plumule lengths acceptable for sale are 2 cm and 1.5 cm, respectively; this minimizes damage in transportation and nursery planting into polybags.

Seeds may be processed in polythene bags and/or trays. Excellent germination results have been achieved using trays in both the hot and the cold rooms (Periasamy *et al.*, 2002). However, there are some disadvantages in using trays, which are as follows:

- The seeds may only be inspected by checking each tray individually, whereas it is easy to observe problems (fungus contamination and/or incorrect moisture) when using polythene bags.
- Controlling seed moisture content is potentially more difficult because moisture loss is more rapid from a tray than from a polythene bag.

However, the above negatives of using trays are compensated for by faster and higher rates of seed germination and by being able to select germinated seed direct from the trays.

9.2   Tools

- Trolley – used to transport stacked seed trays.
- Stackable solid plastic trays – used as seed containers.
- Thermometer – used to measure/monitor room temperatures.
- Thermo-hygrograph – used to measure and monitor room temperature and humidity.
- Heater – used to heat germination rooms to required temperatures.

- Fan – used to blow and spread the heated air evenly around the germination rooms.
- Drying racks or stackable perforated trays – used to air-dry the seed trays.
- Hand sprayer – used to spray distilled water on to seeds to maintain moisture conditions.
- Tooth brush – used to clean away any fungal contamination on the seed surface.
- Plastic bowl – used to contain chemical solutions.
- Cleaning towel – used to dry trays from any remaining wetness from water spraying.
- Label and marker pen – used to record seed identity and location.
- Balance scale – used to weigh discarded seed.
- Net bags – used for soaking batches of seed.
- Plastic container or bags – used for weighing batches of seed.

9.3 Materials

- Stock solutions of fungicides, such as 'Mancozeb' and 'Benomyl', and bactericide streptomycin sulfate in 50 l of water.
- The surface of the plastic trays should be sterilized with 70% ethanol.

Safety equipment

Rubber hand gloves, mask, rubber sandals, goggles and apron.

9.4 Methods

Step 1

Remove the seeds from the cold room (normally a batch comprises about 100,000 seeds for commercial production) and clean with running water. Remove all floating (defective) seeds and record. Place seeds into a net bag and soak in a water tank for 5 days. The water should be changed daily. After 5 days, the seeds are washed in running water.

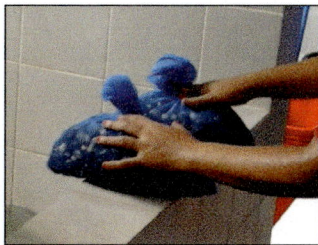

| Seed cleaning | Seed in net bags | Seed soaking |

Step 2

Treat the clean seeds with fungicides by soaking for about 3 min, then move the seeds to drying racks (or stackable perforated trays) for up to 6 h. Samples are taken to check seed viability and seed moisture content (see Chapters 7 and 8 of this manual).

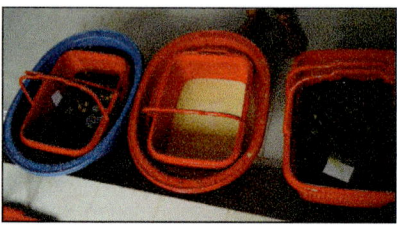

| Soak seed in a solution of sodium hypochlorite | Seed treatment with fungicide solutions | Seed drying |

Step 3

Place dried seeds into stackable plastic trays or plastic bags and transfer to the heating room using a trolley. Heat treatment is required to weaken the operculum before germination: 40–60 days at about 39.5°C. Seeds should be inspected for fungal infection to limit contamination, and water spraying should be carried out every week to maintain the required moisture level.

| Seed inspection | Seed spraying |

Step 4

Remove the seeds from the heating room and clean with running water before starting the second soaking. Floating (defective) seeds are removed and recorded; the remaining seeds are placed into net bags and soaked in water for 3 days. The soaking water should be changed daily. After that, the seeds are cleaned under running water.

Step 5

Treat the clean seeds with fungicides by soaking for about 3 min, then move the seeds to drying racks for up to 30 min until the seeds appear to be 'dull' (>22% moisture content). Samples are taken to check moisture content and seed viability.

| Seed drying | Dry until dull appearance |

Step 6

Place the dried seeds into plastic trays or plastic bags and transfer to the germination room using a trolley. The seeds will start to germinate after 7 days, with the setting temperatures at 26°C (night) – 34°C (day), and can be selected as germinated seeds after plumules and radicles emerge and differentiate. Return the ungerminated seeds to the germination room for up to 60 days. During this period, over 80% of seed should germinate. Periodical water spraying should be carried out to maintain the moisture level required.

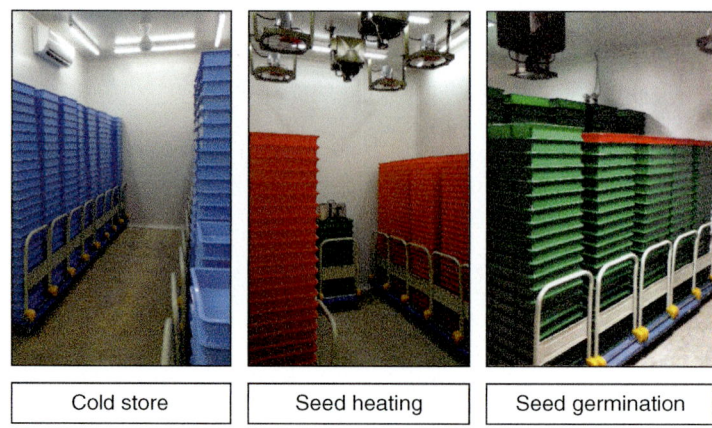

| Cold store | Seed heating | Seed germination |

Step 7

Germinated seeds are inspected for quality and any off-types are discarded. Common off-types are shown below.

Off-type (rejected) seeds

Step 8

Before transferring the germinated seed to the nursery, it may be stored in plastic bags on racks in an air-conditioned room at 19–26°C. Germinated seeds generally remain in good condition for up to 2 weeks. A standard maximum length of the plumule and radicle should be followed to avoid any damage due to overgrown seed during transport to the nursery. Germinated seeds are packed in suitable plastic bags (100/bag) and these are placed into packaging boxes with labels.

| Seed store | Germinated seeds |

9.5 Tenera Purity Testing

As described in Chapters 1, 3 and 4 of this manual, Tenera commercial seed is produced from crossing Dura mother (seed) palms with Pisifera father (pollen) palms. This process is regulated carefully to minimize unwanted pollinations. However, Tenera production is not 100% efficient and non-Tenera seed can occur. As a result, national purity standards are set for planting materials; for example, Indonesia allows up to 2% non-Tenera contaminants and Malaysia up to 5%. Growers are particularly keen to plant high-quality seed, and therefore seed producers adopt quality control measures to guarantee purity levels, e.g. 98% Tenera or higher. This has significant economic implications for growers, as non-Teneras are less productive: Duras have less oil in their fruits compared to Teneras, and Pisiferas are generally fruitless.

Tenera purity can be determined by DNA analysis. Diagnostic markers for Tenera, Dura and Pisifera genotypes are available. These are specific to the shell thickness gene (*Sh*) and have been developed from gene sequence information (Singh *et al.*, 2013) and can differentiate between Tenera, Dura and Pisifera genotypes, *Sh/sh*, *Sh/Sh* and *sh/sh*, respectively. Orion Biosains (www.orionbiosains.com) offers a service to determine Tenera purity from leaf samples of field and nursery palms. Verdant has also developed the technology to test for Tenera purity that can equally be applied to seed samples. Tenera purity of seed samples is expected to become available as a service to breeders, seed producers and growers in the near future. The methods involve DNA extraction from a sample of seed and interrogating the alleles at the *Sh* locus to assess Tenera purity. Services in *Sh* determination may be carried out on a large scale in specialized laboratories that can produce results on a few thousand samples in 1–2 weeks for large-scale seed producers and plantation growers. However, such determinations also have great relevance for small-scale producers, as a few inappropriate palms have a relatively significant effect on stand yield. An issue for smallholders is that they often do not buy seeds or seedlings but instead collect seedlings from good stands of Tenera palms, expecting the seedlings to perform as well as their parents. However, since Teneras are heterozygous at the *Sh* locus (*Sh/sh*), their progeny will segregate roughly as 25% Dura : 50% Tenera : 25% Pisifera, which means only 50% of seedlings from open-pollinated Teneras will be of the desired type. This can be avoided by DNA testing prior to field planting.

References

Alang, Z.C., Moir, G.F.J. and Jones, L.H. (1988) Composition, degradation and utilization of endosperm during germination in the oil palm (*Elaeis guineensis*). *Annals of Botany* 61, 261–268.

Periasamy, A., Gopal, K. and Soh, A.C. (2002) Productivity improvements in seed processing techniques for commercial oil palm seed production. *Planter, Kuala Lumpur* 78, 429–442.

Singh, R., Low, E.T.L., Ooi, L.C.L., Ong-Abdullah, M., Chin, T.N. *et al.* (2013) The oil palm *Shell* gene control oil yield and encodes a homologue of SEEDSTICK. *Nature* 500, 340–344.

Abstract

Seed production processes for breeding follow the same basic processes as described for (Dura) commercial seed. Breeding seed may be either Dura (thick shelled) or Tenera (thin shelled). As in commercial production, oil palm breeding seeds require heat treatment before germination will start. Seed processing procedures described in Chapters 6–9 of this manual should be implemented, but the target is dependent on the numbers of germinated seeds required for breeding purposes, e.g. trialling. Also, Tenera seeds, which are thin shelled, are treated slightly differently to thick-shelled Dura seeds during the heating step to break dormancy. Normally, Tenera seeds have a lower germination rate compared to Duras, due to them being more susceptible to fungal attack.

10.1 General Principles in Breeding Seed Production

Oil palm breeding and methods in crossing are described elsewhere; see, for example, Forster *et al.* (2018) and Setiawati *et al.* (2018), respectively.

Numbers of seed required for breeding

For breeding production, seed processing needs to be adapted to the number of seedlings required for crossing and trial purposes.

Maintain seed quality standards

Avoid and discard seeds damaged during the de-pulping process and remove off-type seeds. Some seeds from breeding come from Tenera females, which have a thin shell, therefore greater care is needed to avoid damage.

© Eddy S. Kelanaputra, Stephen P.C. Nelson, Umi Setiawati, Baihaqi Sitepu, Fazrin Nur, Brian P. Forster and Abdul Razak Purba 2018. *Seed Production in Oil Palm: A Manual*

No fungal contamination

Seed storage and processing rooms must be kept clean and staff must be aware of moisture levels at all times.

Excellent germination and low rejection rates

This requires strict implementation of seed production procedures. Optimum temperatures in the hot room and germination room must be provided consistently. Seed disturbance should be minimized during germination and seed sorting to avoid abnormal germination.

10.2   Breeding Seed Storage

Breeding seed is normally stored in polybags in a cold room at 22°C for up to 1 year; viability will decline with age after a year.

Breeding seed store

10.3   Tools, Materials and Methods

Most of the seed production processes between commercial seeds and breeding seeds are the same with respect to tools, materials and methods, but in some circumstances breeding seeds need different methods, as shown in Table 10.1.

Table 10.1. Table of main differences between commercial and breeding seed production.

Step	Commercial	Breeding
Storage container	Use stackable plastic trays (8 mm thick trays for the cold room and germination room and 6 mm thick for the heating room)	Use plastic bags for Tenera seeds in all processing steps. For Dura seeds, plastic bags are used in the cold room (to maintain high humidity), then use 6 mm thick trays during heating and 8 mm thick trays in the germination room
Soaking 1	3–5 days	5 days
Heating duration	40–60 days	60–80 days
Heating temperature	39.5°C ± 0.5	39.0°C ± 0.5
Germination temperature	26°C (night) – 34°C (day)	Ambient

References

Forster, B.P., Sitepu, B., Setiawati, U., Kelanaputra, E.S., Nur, F., Rusfiandi, H., *et al.* (2018) Oil palm (*Elaeis guineensis*). In: Campos, H. and Caligari, P.D.S. (eds) *Genetic Improvement of Tropical Crops.* Springer International Publishing, Cham, pp. 241–290.

Setiawati, U., Sitepu, B., Nur, F., Forster, B.P. and Dery, S. (2018) *Crossing in Oil Palm: A Manual. Techniques in Plantation Science.* Forster, B.P. and Caligari, P.D.S. (eds). CAB International, Wallingford, UK.

Index

Page numbers in **bold** type refer to figures and tables.

CABI – who we are and what we do

This book is published by **CABI**, an international not-for-profit organisation that improves people's lives worldwide by providing information and applying scientific expertise to solve problems in agriculture and the environment.

CABI is also a global publisher producing key scientific publications, including world renowned databases, as well as compendia, books, ebooks and full text electronic resources. We publish content in a wide range of subject areas including: agriculture and crop science / animal and veterinary sciences / ecology and conservation / environmental science / horticulture and plant sciences / human health, food science and nutrition / international development / leisure and tourism.

The profits from CABI's publishing activities enable us to work with farming communities around the world, supporting them as they battle with poor soil, invasive species and pests and diseases, to improve their livelihoods and help provide food for an ever growing population.

CABI is an international intergovernmental organisation, and we gratefully acknowledge the core financial support from our member countries (and lead agencies) including:

Discover more

To read more about CABI's work, please visit: **www.cabi.org**

Browse our books at: **www.cabi.org/bookshop**,
or explore our online products at: **www.cabi.org/publishing-products**

Interested in writing for CABI? Find our author guidelines here:
www.cabi.org/publishing-products/information-for-authors/